To Agatha—J.W.

**For my talented critique partners, Betsy Snyder,
Lindsay Ward, Kellie DuBay Gillis, and Alissa McGough.
Thank you all!—S.R.**

Rocky Pond Books
An imprint of Penguin Random House LLC
1745 Broadway, New York, New York 10019

First published in the United States of America by Rocky Pond Books,
an imprint of Penguin Random House LLC, 2025

Text copyright © 2025 by Julie Winterbottom

Illustrations copyright © 2025 by Susan Reagan

Visit us online at PenguinRandomHouse.com.

Library of Congress Cataloging-in-Publication Data is available.

ISBN 9780593620229

1 3 5 7 9 10 8 6 4 2

Manufactured in China · TOPL

Design by Cerise Steel · Text set in Andes

The author is grateful to the scientists and historians who shared their knowledge of Ruth Patrick's work and life: Lloyd Ackert,
Don Charles, Ryan Hearty, Sharon Kingsland, Rex Lowe, Robert Peck, and Marina Potapova. Thanks also to Susan Reagan for her
luminous illustrations, to editor Lauri Hornik and designer Cerise Steel, and to Agatha Andrews, Brenda Bowen,
Shannon Savage, and Stephen Wetta.

The illustrator drew visual information and inspiration from the video interview listed in the bibliography, as well as
much other photographic research and her own experience as a frequent visitor to the Clarion River in Pennsylvania. Her
final artwork was created with hand-painted watercolor washes that were combined with Procreate textures
and line, and then finalized in Photoshop.

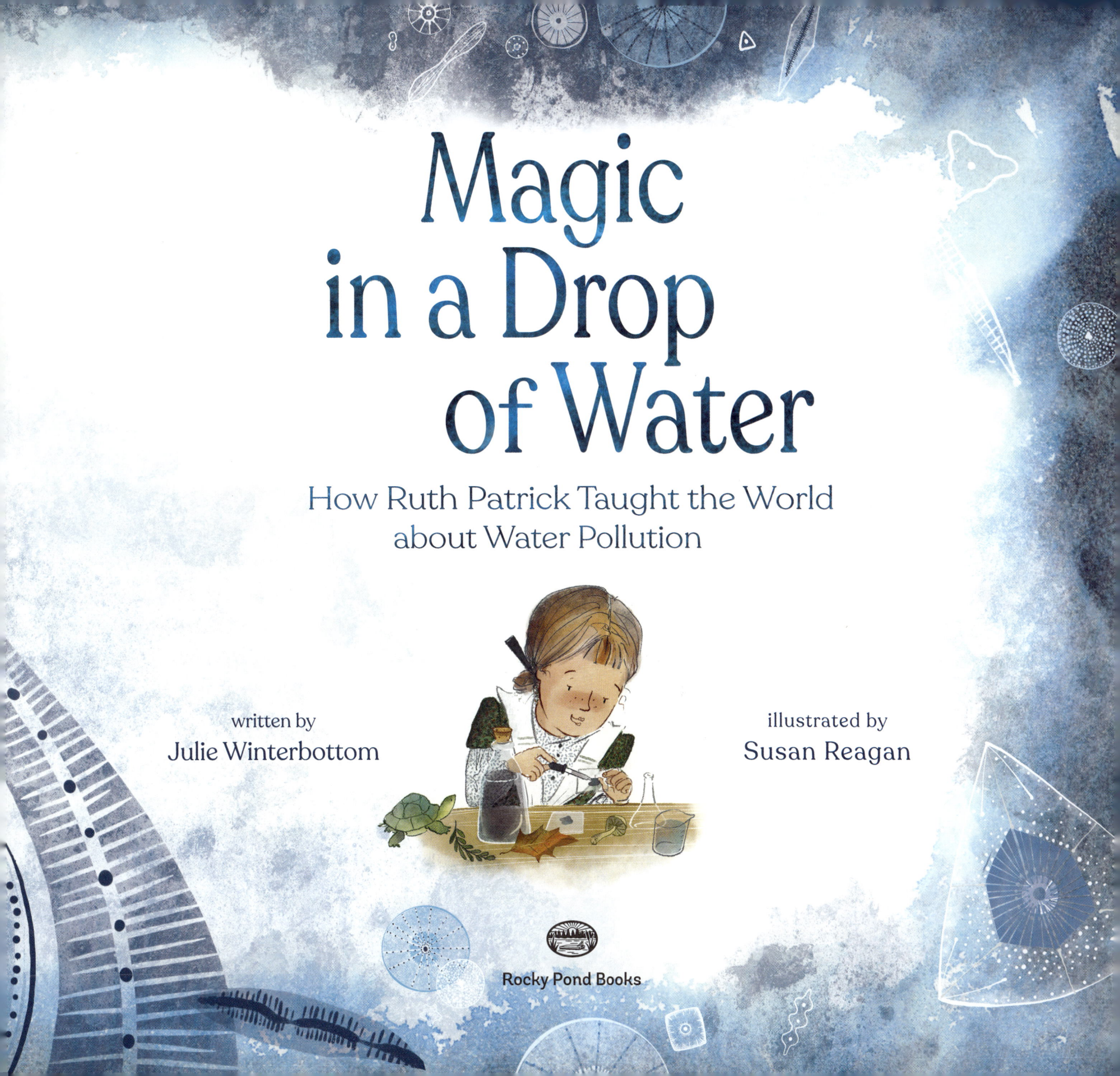

Magic in a Drop of Water

How Ruth Patrick Taught the World about Water Pollution

written by
Julie Winterbottom

illustrated by
Susan Reagan

Rocky Pond Books

When Ruth Patrick was five years old, she fell in love with pond scum.

It happened one Sunday afternoon in 1913. Ruth and her father and sister had just returned from their weekly ramble through the woods near their home in Kansas City, Missouri. Their walks had a mission: Collect anything that looked interesting.

For Ruth, that meant just about everything.

At home, they identified every leaf and leech and mushroom
and fern in their baskets.

One Sunday, they brought home a bottle of slimy brown pond water.
Not the kind of thing you'd imagine falling in love with.

But when Ruth peered at a drop under her father's microscope,
something magical happened.

Jewel-like shapes
glided to and fro,
ovals made of beads,
circles filled with pearls,
shimmering stars and lacy triangles,
each one delicate as a snowflake.

Ruth was entranced.
What were these beautiful gems?
And what were they doing in the pond?

Ruth was looking at diatoms, microscopic algae that live in every body of water on Earth. Each diatom is a single cell protected by glass-like walls. Tiny but mighty, diatoms turn sunlight into food for fish and snails, insects and whales. They produce one-quarter of all the oxygen on Earth. Without them, there would be no life in any pond, stream, river, or ocean. Without them, we would not be here.

Ruth didn't know any of that when she peered through the microscope that day. She just knew she wanted to discover more wonders hiding in watery worlds.

Every chance Ruth got, she waded into swamps and streams, searching for new plants and animals. Was that a fathead minnow tickling her ankle? Was a ghost shrimp hiding in those weeds? Were there diatoms on those slimy rocks?

Her mother was horrified whenever Ruth burst through the door covered in mud, swinging a bucketful of worms and bugs. In the early 1900s girls weren't supposed to get dirty.

"If my friends come by, I wouldn't *think* of saying you were my daughter," she huffed.

Ruth's father, though, was thrilled to see that Ruth loved nature as much as he did. Back then girls were not encouraged to study science. But he thought that was balderdash! He gave Ruth his childhood microscope and she spent hours finding new worlds in a grain of salt, a piece of pollen, a drop of water.

Ruth's father also gave her a tall order. "You must leave the world a better place than you found it," he told his daughters every evening after dinner.

With encouragement from her father, Ruth went to a women's college to study biology and then to graduate school to focus on her true love, diatoms.

She was amazed by how many different species there were. Tens of thousands! Each one was finicky about where it lived. Some could only survive in a bath of salty ocean water. Others only in ice-cold stream water.

Polar nodule

Central nodule

Striae

Raphe

Anatomy of a pennate diatom

One day, she discovered that diatoms could tell stories.

One of Ruth's fellow students wanted to know why, thousands of years ago, all the trees in Virginia's Great Dismal Swamp had died suddenly. He asked Ruth for help. She looked at soil from deep beneath the swamp through her microscope. There she saw ancient diatoms, perfectly preserved by their sturdy glass-like shells. She knew the species. These diatoms lived only in salt water. They told the story:

Long ago,
the ocean flooded the land,
killing all the trees
with its salt.

Ruth wondered, could diatoms tell
other stories about the world?

After graduate school, Ruth got a job at the Academy of Natural Sciences in Philadelphia, Pennsylvania. She kept wading into streams and lakes to learn more about diatoms. She discovered that the tiny algae were even more sensitive than she thought. Some thrived if there were certain chemicals in the water. But others died.

That made Ruth ask, "Could diatoms help with the problem of water pollution?" She had never forgotten her father's words: "You must leave the world a better place." Maybe this was her chance to do that.

Back then, in the 1940s, the nation's rivers were in deep trouble.
New factories were dumping toxic chemicals into the water night
and day. Waste from farms and mines seeped into streams.

Rivers could absorb *some* pollution. But how much was too much?
No one knew. Scientists had no good way to measure the effects of
pollution. They couldn't tell for sure if a body of water was healthy or sick.

Ruth thought maybe diatoms could help. As she pondered the idea,
her mind traveled, like a raft in a river.

She thought about all the plants and animals that live in streams. Not just the diatoms, but the whole community of sponges and spiders, water lilies and worms, everything from the tiniest protozoa to the biggest fish. She realized that if she really wanted to understand pollution, she had to listen to all of them. Not just one voice, but the whole chorus of water dwellers.

No one had ever studied everything in a river before. It would be a huge job.

Ruth jumped in feetfirst.

She brainstormed with her scientist friends. Then she decided. She would identify everything that lived in healthy streams and compare the list to polluted streams. Ruth knew she needed help, so she hired a team of biologists—experts who knew insects and algae and frogs and fish the way she knew diatoms.

She found the perfect place to study: hundreds of miles of streams in Pennsylvania. Some were sparkling clean, others reeked of chemicals. One smelled divine thanks to waste from a chocolate factory.

Then, on a warm June morning in 1948, Ruth led her crew into the water.
They were a strange sight. They juggled buckets and nets and tools for digging and
scraping. They grinned with excitement, thrilled to be part of Ruth's grand experiment.

Every day, Ruth picked a different stream.
One person tested for chemicals while
the others collected samples of
everything that lived there.

Minnows and midges and mussels and mayflies,
protozoa, bryozoa, diatoms, and crane flies.

Northern hogsuckers and pumpkinseed sunfish,
sponges and shiners and spinycheek crayfish.

Riffle beetles, maggots, and wrinkled marshsnails,
tadpoles and darters and European ear snails.

Hundreds and hundreds and hundreds of species.

PROTOZOA

PUMPKINSEED
SUNFISH

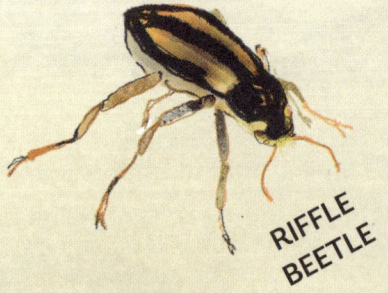

RIFFLE
BEETLE

NORTHERN HOGSUCKER

DIATOMS

BRYOZOA

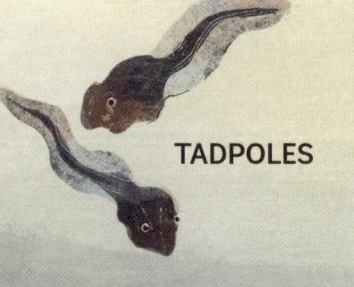

TADPOLES

EUROPEAN
EAR SNAIL

MAYFLY

MIDGE

CRANE FLY

WRINKLED MARSHSNAIL

MINNOW

SHINER

SPINYCHEEK CRAYFISH

DARTER

MUSSELS

SPONGES

MAGGOTS

They worked from dawn until dusk in sweltering heat, Ruth shouting encouragement. "Just one more sample," she would call out as the sun sank in the sky and she bent down to scrape a few more diatoms from a rock.

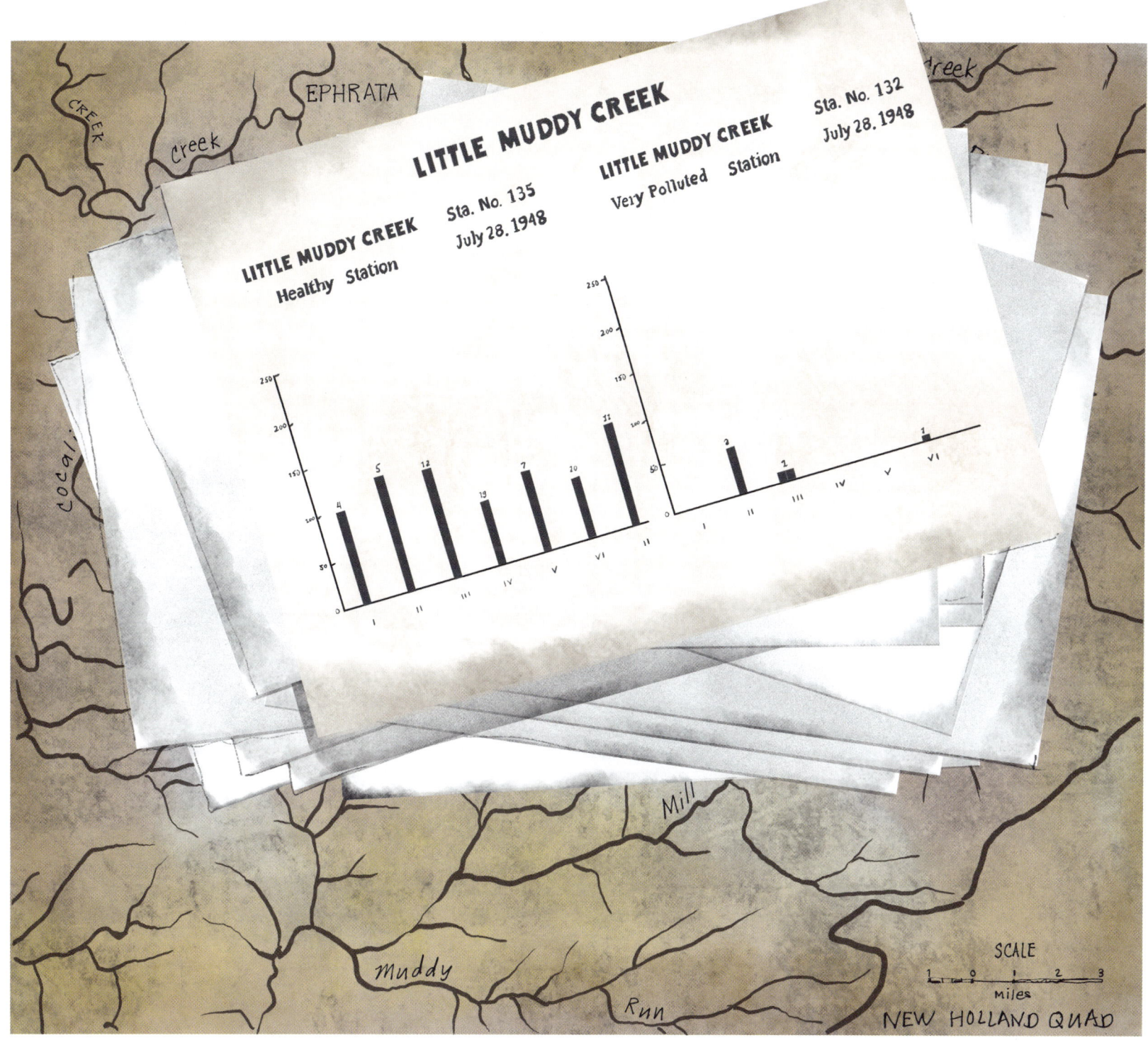

Finally, after a long summer of collecting, they were done. It was time for Ruth to find the story. She studied hundreds of charts and saw a pattern. It was the story of pollution, told by the life in the river.

It went like this:

In healthy water
we are many, many species,
a great variety of creatures,
balanced in number.

No one dominates.

We each play a part in our stable community.

In slightly polluted water,
we are fewer species.
But our numbers are still balanced.

We can all thrive together.

In polluted waters,
the sensitive ones are gone.

Those that remain
grow huge in number.

Our community has lost its balance.

*In very polluted waters,
nothing.*

Ruth had found an ingenious way to measure pollution, by observing the life in a river.
Where there are many different kinds of animals, the water is healthy.
Where there are few—or none—it is polluted.

Today, scientists call this idea biodiversity. They use it to measure pollution everywhere.
Not just in rivers, but in forests and deserts and oceans.
They ask, "Who is here? Who is missing?"

Ruth spent the next sixty years fighting pollution. She showed companies how to stop poisoning rivers and streams. She helped write laws to protect waterways. She taught hundreds of young scientists how to care for the environment. And she never stopped learning herself.

She waded into more than eight hundred rivers and streams all over the world, testing out new ideas about diatoms and pollution. She had some adventures along the way. In South Carolina, venomous snakes dropped from branches overhead as she worked. In Ireland, the British Navy almost shot her, thinking she was trying to blow up their ship.

The problem of water pollution is still here. But now we have the tools to understand and measure it. We know how to listen to the life in a river— or a forest, or our own backyards.

Thanks to the girl who fell in love with the magical world inside a drop of pond scum.

More About Ruth Patrick

When Ruth Patrick led her team of biologists into the streams of the Conestoga Basin in Lancaster County, Pennsylvania, in 1948, she was embarking on an ambitious experiment. What made the endeavor even more remarkable was that a woman was at the helm. In the 1930s and '40s, science was considered a man's world. "Most men in those days believed that women had inferior brains and could not possibly achieve in science," Ruth wrote. Fortunately, Ruth's father and her high school botany teacher had nurtured her passion for science, and that gave her the confidence she needed to pursue a PhD in botany.

When she went to work at the Academy of Natural Sciences of Philadelphia in 1933, she was not made to feel welcome. She was the only female scientist on staff and she worked without pay until 1945, despite being a world expert on diatoms (pronounced DYE-uh-toms). She was told not to wear lipstick because it made her seem frivolous. And when a funder offered the Academy a large sum of money for Ruth to study river pollution, the Academy president replied that she could not possibly be in charge because "all women waste money."

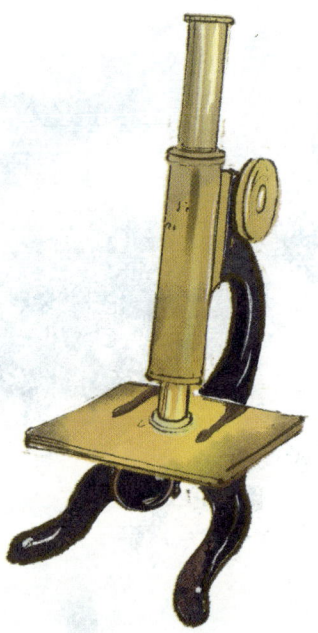

Ruth refused to let the prejudice she encountered keep her from doing the work she loved. After completing her landmark 1948 study, she found mentors and funding outside the Academy to continue her research in rivers around the world. She confirmed her breakthrough finding that biodiversity is a measure of environmental health, now known as the Patrick Principle. She also showed that diatoms are excellent indicators of water pollution. She even built an artificial stream to learn how ecosystems recover from damage. In 1996, she was given the National Medal of Science, the nation's highest science award, for her pioneering research.

Ruth also spent countless hours persuading politicians and company executives to do something about water pollution. She helped write the 1972 Clean Water Act, the nation's most important water protection law, and was one of the first scientists to speak about global warming. She taught younger scientists—many of them women—how to study a stream holistically. And she never lost her deep curiosity about the natural world. At age one hundred, she was still donning her white pith helmet and wading into streams to look for diatoms. She was still asking herself—and everyone she met—her favorite question: What have you learned today?

Sources for Quotes

"If my friends . . ." Patrick, "Hometown Legends," 4:23. • "You must . . ." Patrick, "Acceptance Speech for 1996 Lifetime Achievement Award" ASLO Bulletin, Summer 1996, p. 12. • "Just one more . . ." Untitled memoir by John Cairns, Jr, 2014, p. 28. (vtechworks.lib.vt.edu/bitstream/handle/10919/25016/CairnsAutobiography.pdf?sequence=1) • "Most men . . ." Patrick, "The Development of the Science of Aquatic Ecosystems," p. 6. • "all women . . ." Patrick, "Hometown Legends," 15:22.

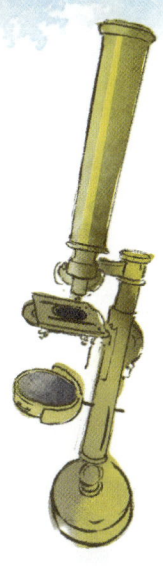

Timeline

November 26, 1907: Born in Topeka, Kansas

1929: Graduates from Coker College with a degree in biology

1931: Marries Charles Hodge IV, an entomologist

1933: Moves to Philadelphia and begins working at Academy of Natural Sciences of Philadelphia (ANSP)

1934: Receives PhD in botany from the University of Virginia

1947: Founds Limnology Department (now the Patrick Center for Environmental Research) at ANSP to study pollution in streams and rivers

1948: Directs landmark study of the Conestoga River Basin

1950: Begins teaching botany at the University of Pennsylvania

1951: Son Charles is born

1954: Invents the diatometer, a device for collecting diatoms to assess water pollution

1955: Leads expedition to Amazon River

1966: Publishes study of the effects of a nuclear plant on the Savannah River

1972: Clean Water Act, which Ruth helped write, becomes law

1996: Awarded National Medal of Science, nation's highest science honor

September 23, 2013: Dies at age 105

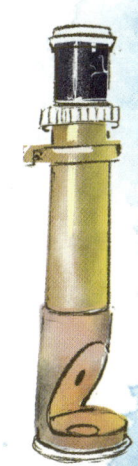

Selected Bibliography

Hearty, Ryan. "Redefining Boundaries: Ruth Myrtle Patrick's Ecological Program at the Academy of Natural Sciences of Philadelphia, 1947–1975." *Journal of the History of Biology*, vol. 53, 2020, pp. 587–630.

"Hometown Legends: Ruth Patrick." WHYY-TV. video.whyy.org/video/whyy-specials-ruth-patrick/

Patrick, Ruth. "The Development of the Science of Aquatic Ecosystems." *Annual Review of Energy and the Environment*, vol. 22, no. 1, 1997, pp. 1–11.

Patrick, Ruth. "Water Pollution." *Life Stories: World-Renowned Scientists Reflect on their Lives and the Future of Life on Earth*, edited by Heather Newbold, Berkeley, University of California Press, 2000, pp. 85–92.

Steinmann, Marion. "Rivers of America: the Source Is Ruth Patrick." *RF Illustrated*, June 1983, pp. 14–16.

Swaby, Rachel. *Headstrong: 52 Women Who Changed Science–and the World*, New York, Broadway Books, 2015.